D0949885

LITTLE BOOK OF PIGS

WEIDENFELD & NICOLSON
LONDON

Sumptuously Shapely

I never could imagine why pigs should not be kept as pets. To begin with, pigs are very beautiful animals. Those who think otherwise are those who do not look at anything with their own eyes, but only through other people's eyeglasses. The actual lines of a pig (I mean a really fat pig) are among the loveliest and most luxuriant in nature; the pig has the same great curves, swift and yet heavy, which we see in rushing water or in rolling cloud . . .

Now, there is no point of view from which a really corpulent pig is not full of sumptuous and satisfying curves . . .

In short, he has that fuller, subtler, and more universal kind of shapeliness which the unthinking (gazing at pigs and distinguished journalists) mistake for a mere absence of shape.

THE USES OF DIVERSITY

Gilbert Keith Chesterton 1874-1936

THIS REMARKABLE ANIMAL
Benjamin Gale 19th century

CIRCE
Wright Barker early 20th century

There Was a Lady Loved a Swine

There was a lady loved
a swine,
'Honey,' quoth she,
'Pig-Hog, wilt thou be mine?'
'Humph!' quoth he.

'I'll build thee a silver sty,
Honey,' quoth she,
'And in it thou shalt lie.'
'Humph!' quoth he.

'Pinned with a silver pin,
Honey,' quoth she,
'That thou may'st go out
and in.'
'Humph!' quoth he.

'Wilt thou have me now,
Honey?' quoth she.
'Speak, or my heart will break.'
'Humph!' quoth he.

TRADITIONAL

Boar Play

'And is Tom savage?' asked Mrs Bradley, looking down on the boar.

'Oh, no, not a bit, except with strangers. Look.' He opened the gate and walked in. Tom backed away, stood with his back to the fencing, and scratched the ground with his feet.

'Come on, you old stupid,' said Carey. Slowly the boar advanced, as delicately as though he were treading a minuet, but when he was less than four feet away from Carey he made a frenzied rush. Carey leapt aside like a Spanish bullfighter, slapped the boar on the hams and faced him again. This happened three times, and then the boar turned quiet and walked away. Carey went after him, held his head, and showed Mrs Bradley his tusks. Then he walked calmly out of the run, and fastened the door behind him, wiping his hands down his trousers.

'I shouldn't like to say he wasn't savage,' was Mrs Bradley's comment.

DEAD MEN'S MORRIS

Gladys Mitchell 1901-1983

WILD BOAR
AND WOLF
F. Gauermann

FORAGING FOR NUTS

Francis Danby 1793-1861

The Company he Chooses

'Twas an evening in November,
As I very well remember,
I was strolling down the street in drunken pride,
But my knees were all a'flutter
So I landed in the gutter,
And a pig came up and lay down by my side.
Yes I lay there in the gutter
Thinking thoughts I could not utter,
When a colleen passing by did softly say,
'Ye can tell a man that boozes
By the company he chooses' -
At that, the pig got up and
walked away!

TRADITIONAL

A Grunting Baby

THE BABY grunted again, and Alice looked very anxiously into its face to see what was the matter with it. There could be no doubt that it had a very turn-up nose, much more like a snout than a real nose; also its eyes were getting extremely small for a baby; altogether Alice did not like the look of the thing at all. 'But perhaps it was only sobbing,' she thought, and looked into its eyes again to see if there were any tears.

No, there were no tears. 'If you're going to turn into a pig, my dear,' said Alice, seriously, 'I'll have nothing more to do with you. Mind now!' The poor little thing sobbed again (or grunted, it was impossible to say which) and they went on for some while in silence. Alice was just beginning to think to herself, 'Now, what am I to do with this creature when I get it home?' when it grunted again, so violently, that she looked down into its face in some alarm. This time there could be no mistake about it; it was neither more nor less than a pig and she felt that it would be quite absurd for her to carry it any further. So she set the little creature down and felt quite relieved to see it trot away quietly into the wood.

ALICE'S ADVENTURES IN WONDERLAND

Lewis Carroll 1832-1898

STUDY OF A FIERCE BOAR IN THE FOREST

Theodor Julius Kiellerup 19th century

PIGS

'Do look at those pigs as they lie in the straw,'
Said Dick to his father one day;
'They keep eating longer than I ever saw,
What nasty fat gluttons are they.'

'I see they are feasting' his father replied,
'They eat a great deal, I allow;
But let us remember, before we deride,
'Tis the nature, my dear, of a sow.

'But when a great boy, such as you, my dear Dick,
Does nothing but eat all the day,
And keeps sucking good things till he makes himself sick,
What a glutton! Indeed, we may say.

'When plumcake and sugar for ever he picks,
And sweetmeats, and comfits, and figs;
Pray let him get rid of his own nasty tricks,
And then he may laugh at the pigs.'

Jane Taylor 1783-1824

FATHER AND SON ADMIRE THE PIGS
Catherine Bell 19th century

HEY DIDDLE DIDDLE
Randolph Caldecott 1846-1886

Performing Pigs

IT'S MIDSUMMER EVE tomorrow,' said Badger. 'Well, I've never seen a theatre, so we will go together, and see the fun, whatever it is' . . .

One part of the stable was screened off with a leafy curtain and from behind it came muffled laughter, high squeals, and subdued whispers . . .

Sally the mare twitched the curtain aside, and nibbled a few leaves in her excitement. Everybody cried 'Oh-o-o-o-o-o-o!'

There were the seven little pigs from the pig-cote, dressed as fairies, in pink skirts with wreaths of rosebuds round their pink ears. They danced on their nimble black toes, and swung their ballet skirts. They pirouetted until the hens cried out to them. A band of music makers played in a corner . . .

'Come along, Sam, and join us,' they beckoned, and Sam stepped shyly through the little dancing pigs who never stopped whirling. He tuned his fiddle and sat down in the corner. Soon he was sawing with might and main, trying to keep time with the squealing of the Scottie's bagpipes, the fluting of the lamb's pipe and the drumming of the little cat.

ADVENTURES OF SAM PIG *Alison Uttley* 1884-1976

The Boar Hunt Toast

PLEDGE me next the glorious chase
When the mighty boars ahead,
He, the noblest of the race,
In the mountain jungle bred
Swifter than the slender deer,
Bounding over Deccan's plain
Who can stay his proud career,
Who can hope his tusks to gain?

TRADITIONAL

BOAR HUNTING
Francisco Goya
1746-1828

PRIZED PIGS

During its lifetime the pig was an important member of the family, and its health and condition were regularly reported in letters to children away from home, together with news of their brothers and sisters. Men callers on Sunday afternoons came, not to see the family, but the pig, and would lounge with its owner against the pigsty door for an hour, scratching piggy's back and praising his points or turning up their noses in criticism. Ten to fifteen shillings was the price paid for a pigling when weaned, and they all delighted in getting a bargain. Some men swore by the 'dilling', as the smallest of a litter was called, saying it was little and good, and would soon catch up; others preferred to give a few shillings more for a larger young pig.

LARK RISE
TO CANDLEFORD
Flora Thompson
1876-1947

TIRED OUT

Ellen Lucas 19th century

To Market

To market, to market,
To buy a fat pig.
Home again, home again,
Jiggety jig!

To market, to market,
To buy a fat hog.
Home again, home again,
Jiggety jog!

TRADITIONAL

MORNING HIGGLERS PREPARING FOR MARKET *George Morland* 18th century

The Cleverest Of Animals

THE work of teaching and organizing fell naturally upon the pigs, who were generally recognized as being the cleverest of animals. Pre-eminent among the pigs were two young boars named Snowball and Napoleon, whom Mr Jones was breeding for sale. Napoleon was a large, rather fierce-looking Berkshire boar, the only Berkshire on the farm, not much of a talker, but with a reputation for getting his own way. Snowball was a more vivacious pig than Napoleon, quicker in speech and more inventive, but was not considered to have the same depth of character. All the other male pigs on the farm were porkers. The best known among them was a small fat pig named Squealer, with very round cheeks, twinkling eyes, nimble movements, and a shrill voice. He was a brilliant talker, and when he was arguing some difficult point he had a way of skipping from side to side and whisking his tail which was somehow very persuasive.

ANIMAL FARM *George Orwell* 1903-1950

THE MARCH PAST
William Weekes 19th century

COLLECTING ACORNS FOR PIGS:
NOVEMBER LABOUR
The Playfair Book of Hours 15th century

Falling Food

No more the fields with scattered grain supply
The restless wandering tenants of the sty;
From oak to oak they run with eager haste,
And wrangling share the first delicious taste
Of fallen acorns; yet but thinly found
Till the strong gale has shook them to the ground.
It comes; and the roaring woods obedient wave:
Their home well-pleased the joint adventurers leave:
The trudging sow leads forth her numerous young,
Playful, and white, and clean, the briars among,
Till briars and thorns increasing, fence them round,
Where last year's mouldering leaves bestrew the ground,
And o'er their heads, loud lash'd by furious squalls,
Bright from their cups the rattling treasure falls;
Hot, thirsty food; whence doubly sweet and cool
The welcome margin of some rush-grown pool.

THE FARMER'S BOY
Robert Bloomfield 1766-1823

A Sagacious Sow

THE NATURAL TERM of a hog's life is little known, and the reason is plain - because it is neither profitable nor convenient to keep that turbulent animal to the full extent of its times: however, my neighbour, a man of substance, who had no occasion to study every little advantage to a nicety, kept a half-bred bantam-sow, who was as thick as she was long, and whose belly swept on the ground, till she was advanced to her seventeenth year, at which period she showed some tokens of age by the decay of her teeth and the decline of her fertility.

For about ten years this prolific mother produced two litters in the year of about ten at a time, and once above twenty at a litter; but, as there were near double the number of pigs to that of teats many died. From long experience in the world this female was grown very sagacious and artful. When she found occasion to converse with a boar she used to open all the intervening gates, and march, by herself, up to a distant farm where one was kept; and when her purpose was served would return by the same means.

NATURAL HISTORY OF SELBORNE

Gilbert White 1720-1793

FARMYARD SCENE WITH
PIGS, DUCKS & POULTRY
John Frederick Herring 19th century

Sleeping Swine

THE SUNNY slow lulling afternoon yawns and moons through the dozy town. The sea lolls, laps and idles in, with fishes sleeping in its lap. The meadows still as Sunday, the shut-eye tasselled bulls, the goat-and-daisy dingles, nap happy and lazy. The dumb duckponds snooze. Clouds sag and pillow on Llaregyb Hill. Pigs grunt in a wet wallow-bath, and smile as they snort and dream. They dream of the acorned swill of the world, the rooting for pig-fruit, the bagpipe dugs of the mother sow, the squeal and snuffle of yesses of the women pigs in rut. They mud-bask and snout in the pig-loving sun; their tails curl; they rollick and slobber and snore to deep, smug, after-swill sleep. Donkeys angelically drowse on Donkey Down.

UNDER MILK WOOD
Dylan Thomas 1914-1953

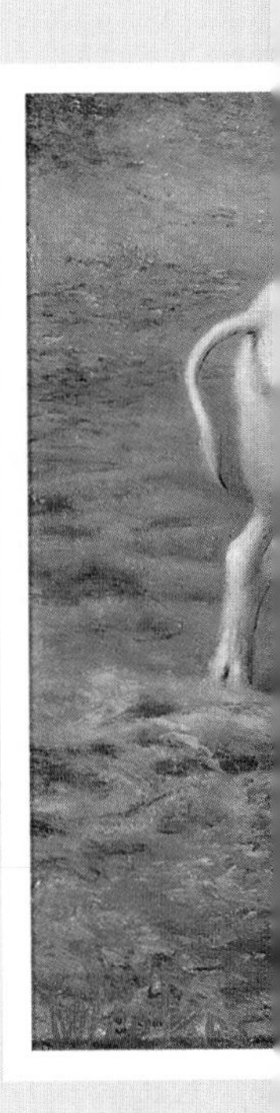

AN UNAPPRECIATIVE AUDIENCE
William Weekes 19th century

The Poor Man's Pig

Already fallen plum-bloom stars the green
And apple-boughs as knarred as old toad's backs
Wear their small roses ere a rose is seen;
The building thrush watches old Job who stacks
The bright-peeled osiers on the sunny fence,
The pent sow grunts to hear him stumping by,
And tries to push the bolt and scamper thence,
But her ringed snout still keeps her to the sty.

Then out he lets her run; away she snorts
In bundling gallop for the cottage door,
With hungry hubbub begging crusts and orts,
Then like a whirlwind bumping round once more;
Nuzzling the dog, making the pullets run,
And sulky as a child when her play's done.

Edmund Blunden 1896-1974

THE RIDE
William Bromley
19th century

PIGS IN THE WOODS

19th century

The Pigs & The Charcoal Burner

The old Pig said to the little pigs,
'In the forest is truffles and mast,
Follow me then, all ye little pigs,
Follow me fast!'

Walter de la Mare 1873-1956

A DIRGE

LITTLE BETTY WINKLE she had a pig,
It was a little pig not very big;
When he was alive he liv'd in clover,
But now he's dead and that's all over;
Johnny Winkle he
Sat down and cried,
Betty Winkle she
Laid down and died;
So there was an end of one, two and three,
Johnny Winkle he,
Betty Winkle she,
And Piggy Wiggie.

TRADITIONAL

CHILDREN & PIGS
Edith Berkeley
19th century

A BROOD SOW

Thomas Landseer 1795-1880

There was a Pig

There was a pig, that sat alone,
Beside a ruined pump.
By day and night he made his moan:
It would have stirred a heart of stone
To see him wring his hoofs and groan,
Because he could not jump.

Lewis Carroll 1832-1898

The Temptation of St Anthony

Goblins came, on mischief bent,
To Saint Anthony in Lent.
'Come, ye goblins, small and big,
We will kill the hermit's pig.
While the good monk minds
his book,
We, the hams will cure and cook.
While he goes down on his knees,
We will fry the sausages.
While he on his breast doth beat,
We will grill the tender feet.
While he David's Psalms
doth sing,
We will all to table bring.'
On his knees went Anthony
To those imps of Barbary.
'Good, kind goblins, spare
his life,
He to me is child and wife.
He indeed is good and mild
As 'twere any chrisom child.
He is my felicity,
Spare, oh spare my pig to me!'
But the pig they did not spare,
Did not heed the hermit's prayer.
They the hams did cure
and cook,
Still the good Saint read his book.
When they fried the sausages
Still he rose not from his knees.
When they grilled the tender feet
He ceased not his breast to beat.
They did all to table bring,
He for grace the Psalms did sing.
All at once the morning broke,
From his dream the monk awoke.
There in the kind light of day
Was the little pig at play.

TRADITIONAL
FRENCH CHANSON

KILLING THE BOAR

Bedford Hours 15th century

QUIET STABLE

John Frederick Herring 1795-1865

As I Looked Out

As I looked out on Saturday last,
A fat little pig went hurrying past.
Over his shoulders he wore a shawl,
Although he didn't seem cold at all.
I waved at him, but he didn't see,
For he never so much as looked at me.

Once again, when the moon was high,
I saw the little pig hurrying by;
Back he came at a terrible pace,
The moonlight shone on his little pink face,
And he smiled with a face that was quite content.
But never I knew where that little pig went.

TRADITIONAL

A HAPPY COUPLE

early 20th century

THE PORTLY PATRICIA

WELL, to go back to Patricia Portly . . .

She lay awake many nights trying to think up a new scent to be used at her wedding. You see she wanted a nosegay and - '

'A nosebag, you mean,' Jip put in with another grunt.

'She wanted a nosegay,' Gub-Gub went on, 'which would be something that had never been used by a bride before. And finally she decided on a bunch of Italian Forget-me-nots.'

'I never heard of the flower,' said Too-Too.

'Well, they're not exactly flowers,' said Gub-Gub. 'They are those long green onions that come in the spring. You see the very refined pig society in which Patricia moved did not call them onions. So they changed the name to Italian Forget-me-nots. They became very fashionable after that and were nearly always used at pig weddings.'

GUB-GUB'S BOOK
Hugh Lofting 1886-1947

THE WONDERFUL PIG
18th century

LITERARY LEANINGS

O heavy day! Oh day
of woe!
To misery a poster,
Why was I ever
farrow'd - why
Not spitted for
a roaster?

In this world, pigs,
as well as men,
Must dance to
fortune's fiddlings,
But must I give
the classics up,
For barley-meal
and middlings?

Of what avail that I
could spell
And read, just like
my betters,
If I must come to
this at last,
To litters, not to letters?

O, why are pigs made
scholars of?
It baffles my discerning,
What griskins, fry,
and chitterlings
Can have to do
with learning.

Alas! my learning once
drew cash,
But public fame's
unstable,
So I must turn a
pig again,
And fatten for the table.

THE LAMENT OF TOBY
THE LEARNED PIG
Thomas Hood 1799-1824

One of the Litter

I FINISHED drying my arms and was about to make a casual reference to the kitten when Mr Butler handed me my jacket.

'Come round here with me if you've got a minute,' he said. 'I've got summat to show you' . . .

I stared unbelievingly down at a large sow stretched comfortably on her side, suckling a litter of about twelve piglets and right in the middle of the long pink row, furry black and incongruous, was Moses. He had a teat in his mouth and was absorbing his nourishment with the same rapt enjoyment as his smooth-skinned fellows on either side . . .

Moses for his part appeared to find the society of pigs very congenial. When the piglets curled up together and settled down for a sleep Moses would be somewhere in the heap and when his young colleagues were weaned at eight weeks he showed his attachment to Bertha by spending most of his time with her.

VETS IN HARNESS
James Herriot

THE FARMYARD

John Frederick Herring 1795-1865

Jack Sprat's Pig

Little Jack Sprat
Once had a pig;
It was not very little,
Nor yet very big,
It was not very lean,
It was not very fat -
It's a good pig to grunt,
Said little Jack Sprat.

TRADITIONAL

EFFECTS OF GOOD GOVERNMENT
IN THE COUNTRY
Ambrogio Lorenzetti 14th century

THE INTRUDER
William Weekes 19th century

DESIGNED FOR DINING

As for greed, certainly not even the most sincere apologist of pigs or lover of bacon can deny that they enjoy their victuals. But reflect, reader, how it would be with you if you had an immensely long, barrel-shaped and capacious body carried on four very short legs: if you had a nose (or snout) especially constructed and designed to go to the root of matters: if you had a mouth of peculiar capacity, stretching almost from ear to ear. (And, by the way, what charming ears, too, eminently adapted for flapping and, at the same time, for composing the eye for slumber beneath their ample shade!)

Would you not enjoy your food even more than you do now? Would you not grunt, and even slightly squeal, with the excruciating ecstasy of creamy, rich barley-meal, as it entered your long and wide mouth, gurgled in your roomy throat and flowed on into that vast stomach forever clamouring to be soothed?

STORM AND PEACE

John Beresford 1873-1947

PROBLEMS FOR PIGLETS

By and by Piglet woke up. As soon as he woke he said to himself, 'Oh!' Then he said bravely, 'Yes,' and then, still more bravely, 'Quite so.' But he didn't feel very brave, for the word which was really jiggeting about in his brain was 'Heffalumps.'

What was a Heffalump like?

Was it Fierce?

Did it come when you whistled? And *how* did it come?

Was it Fond of Pigs at all?

If it was Fond of Pigs, did it make any difference *what sort of Pig?*

Supposing it was Fierce with Pigs, would it make any difference *if the Pig had a grandfather called TRESPASSERS WILLIAM?*

He didn't know the answer to any of these questions . . . and he was going to see his first Heffalump in about an hour from now!

Of course Pooh would be with him, and it was much more friendly with two. But suppose Heffalumps were Very Fierce with Pigs *and* Bears? Wouldn't it be better to pretend that he had a headache and couldn't go up to the Six Pine Trees this morning?

WINNIE-THE-POOH

Alan Alexander Milne 1882-1956

WILD BOAR

early 20th century

Whose Little Pigs?

Whose little pigs are these, these, these?
Whose little pigs are these?
They are Roger the Cook's, I know
by their looks;
I found them among my peas.

Go pound them, go pound them.
I dare not on my life,
For though I love not Roger the Cook,
I dearly love his wife.

TRADITIONAL

NOVEMBER
Limbourg Brothers
15th century

Acknowledgements

First published in Great Britain in 1993 by George Weidenfeld and Nicolson Ltd
Orion House, 5 Upper St Martin's Lane, London WC2H 9EA

British Library Cataloguing in Publication Data. A catalogue record for this book is available from the British Library.

Designed and edited by
THE BRIDGEWATER BOOK COMPANY
Words and Pictures chosen by RHODA NOTTRIDGE
Typesetting by VANESSA GOOD
Printed in Italy

The publishers wish to thank the following for the use of pictures:
THE BRIDGEMAN ART LIBRARY: front cover and pages 3, 36, 39, 55; E.T.ARCHIVE: back cover and pages 7, 14, 17, 20, 24, 49; MARY EVANS PICTURE LIBRARY: pages 13, 32, 35, 42, 44, 53; FINE ART PHOTOGRAPHS: page 4, 8, 11, 19, 23, 27, 29, 31, 40, 47, 50.

The publishers gratefully acknowledge permission to reproduce the following material in this book:
p.6 *Dead Men's Morris* by Gladys Mitchell, Michael Joseph.
p.15 *The Adventures of Sam Pig* by AlisonUttley, Faber & Faber Ltd.
p. 18 *Lark Rise To Candleford* by Flora Thompson, 1945 by permission of Oxford University Press.
p.22 *Animal Farm* by George Orwell by permission the estate of the late Sonia Brownell Orwell and Martin Secker & Warburg Ltd and Harcourt, Brace, Jovanovich, USA.
p.28 *Under Milk Wood* by Dylan Thomas (J.M. Dent and Sons Ltd.) by permission David Higham Associates.
p.30 *The Poor Man's Pig* by Edmund Blunden reprinted by permission of the Peters, Fraser & Dunlop Group Ltd.
p.33 *Extract from The Pigs & The Charcoal Burner* by Walter de la Mare by permission The Literary Trustees of Walter de la Mare and The Society of Authors as their representative.
p.43 *Gub-Gub's Book* by Hugh Lofting; Wyman, Bautzer, Kuchell, Silbert.
p.46 *Vet in Harness* by James Herriot by permission David Higham Associates and St Martins Press (USA).
p.51 *Storm and Peace* (Reflections on Bacon) by John Beresford by permission The Spectator.
p.52 *Winnie-the-Pooh* by A.A.Milne published by Methuen Childrens Books, copyright renewal ©1954 A.A. Milne.

Every effort has been made to trace all copyright holders and obtain permissions. The editor and publishers sincerely apologise for any inadvertent errors or omissions and will be happy to correct them in any future edition.